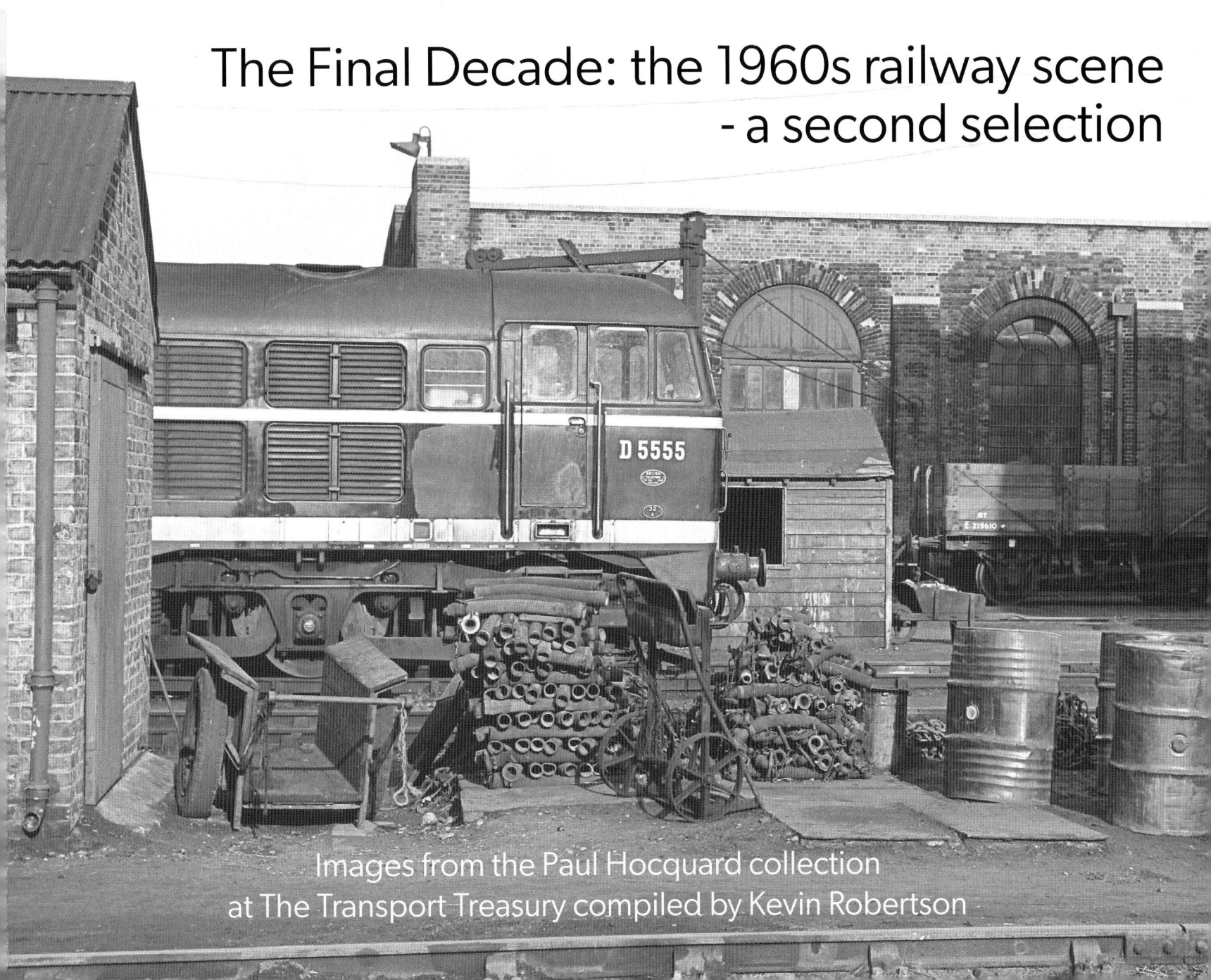

© Images and design: The Transport Treasury 2023. Text: Kevin Robertson

ISBN 978-1-913893-34-7

First Published in 2023 by Transport Treasury Publishing Ltd. 16 Highworth Close, High Wycombe, HP13 7PJ
Totem Publishing, an imprint of Transport Treasury Publishing.

The copyright holders hereby give notice that all rights to this work are reserved.
Aside from brief passages for the purpose of review, no part of this work may be reproduced,
copied by electronic or other means, or otherwise stored in any information storage and
retrieval system without written permission from the Publisher. This includes the illustrations
herein which shall remain the copyright of the copyright holder.

www.ttpublishing.co.uk

Printed in Taxien, Malta by Gutenberg Press Ltd.

Front cover - Where once there would have been steam....? Diesels in the environment of a steam depot. Modern traction, well at the time modern but today considered almost as historic as the steam engines they replaced.

Frontispiece - The clutter associated with the railway for decades and now with the first glimpes of the revolution that was to transform the railways from the late 1950s onwards. Modernisation, announced in 1955, was to sweep away steam in favour of diesel and electric although naturally there would be a period of transition - and which of course is the subject of this book. In this view is Brush Type 2, No D5555; in later years to be designated as a 'Class 31' but that was still some time in the future. Instead for the present we have the relics of the old railway in the form of various spares, detritus perhaps but it has to be said for the time, a relatively neat display. As with Paul's view, no date or location is given but we may be certain it is Eastern Region.

Rear cover - Through Southern to Western service. The Reading to Redhill line; a cross-country route on which steam survived perhaps longer than it should.

'The Final Decade: the 1960s railway scene - a second selection' is one of a series of books on specialist transport subjects
published in strictly limited numbers and produced under the Totem Publishing imprint using material only available at
The Transport Treasury

Introduction

In commencing this second book in the series of Paul Hocquard's wonderful images I am grateful to all who clearly shared my own view on this little known but master of photography.

Speaking personally, it was many years before I realised how a seemingly ordinary scene might be transformed by the clever inclusion of nearby attributes; items such as signals, buildings and above all people combining to give 'life' to what could otherwise be a bland subject.

Once more and speaking from a personal perspective, I had previously made a point of looking for a 'clean' area around the subject matter; but we all learn.

Whilst it was certainly a pleasure to be invited by Transport Treasury to compile this second volume it remains a great pity that despite the first book of Paul Hocquard material having sold out, we are still no further in learning more about the photographer himself. In consequence we remain ignorant of dates and on occasions locations.

Whatever, on behalf of the publisher and myself, I do hope this second selection will rekindle as many memories as with the first book. As before no attempt has been made to group or 'regionalise' the order, instead we are content to let the pictures speak for themselves.

Kevin Robertson. West Berkshire, 2023.

In the grey mists of the land, an unidentified light engine, J39 perhaps, sets off across the first reaches of the Tay Bridge heading south. It was here in the last days of 1879 the bridge collapsed during a gale taking with it a passenger train, there were no survivors. The story is told in gripping detail in John Prebble's excellent book 'The High Girders'. We may be sure the disaster crossed the minds of some as they ventured over the replacement structure in later years.

So where to start, and the answer has to be where we left off! Here Paul has captured the character of a virtually unchanged station building from the days of the London and Southampton railway and dating back close to 200 years. In case the wording is not easily read, it is Micheldever between Basingstoke and Winchester and we are looking at the up side. The all-round canopy is original although naturally some additions have taken place - and more recently some removals as well. Notwithstanding the changes it remains virtually unaltered today.

How many youngsters were railway spotters in the 1950s and 1960s? Likely the total was in the hundreds of thousands with most possessing an Ian Allan book relative to the region - or if they were really well disposed, the combined volume of all regions. These two are on the platform at Birmingham Snow Hill; the short trousers and long socks may be recalled. No bottle of Tizer or sandwiches so either consumed already or perhaps a short visit. We will be meeting them again later in this work.

The now closed station at Saffron Walden had been on the short connecting line between Audley End and Bartlow. Opened in 1865, for its first year it served as a terminus until the remaining section to Bartlow as complete. As an independent concern the fortunes of the railway company were not good - something so many small independent concerns would discover - and in this case the line was sold to the Great Eastern Railway in 1877. It subsequently passed to the LNER and then British Railways in 1948. The 1950s brought about a severe decline and despite the use of rail-buses there was no escaping the inevitable closure to passengers in September 1964.

This view from the station platform is looking towards Audley End. Today an access road to housing on the station site runs along the trackbed, the main station building survives, also having been converted to residential accommodation.

Great North of Scotland Railway No 49 was one of batch of engines of the F class designed by Thomas Heywood in his role as Locomotive Superintendent of the GNoSR.

No 49 entered service in October 1920 having been built by the North British Locomotive Company and given the name *Gordon Highlander*. In LNER days it carried the numbers 6849 and then 2277 and finally under BR No 62277.

Replaced in their duties by larger and newer B1 class 4-6-0 engines, withdrawals had commenced in 1946, No 62249 the last working member of the class withdrawn from Keith in June 1958. Amongst its last duties were services between Elgin and Banff when it was noted to be in very poor external condition the only clean part of the engine being its nameplate.

All this was to change as 'No 49' had been earmarked for preservation as part of the National collection and in consequence was taken to Inverurie Works and restored. Externally it was given the pre-Heywood GNoSR green livery although in practice this had never been carried.

In its restored condition it was based at Dawsholm shed, Glasgow and was a popular choice for special trains over a number of years until placed as a static exhibit in the Glasgow Transport Museum. The engine was subsequently moved on loan to the Museum of Scottish Railways at Bo'ness.

Servicing No 49 whilst on special duty.

As an example of its travels, in 1959 the engine was reported to have been used on at least eight special workings, including for the Scottish Industrial Exhibition and the Golden Jubilee special of the Stephenson Locomotive Society. Sometimes these workings also included other preserved and restored locomotives such as No 103, the Jones Goods, Caledonian Single No 123, the North British 4-4-0 *Glen Douglas*, and even No 3440 *City of Truro*.

The boy on the right is calmly watching proceedings oblivious perhaps to the soaking he might soon receive if the bag of the water column is released whilst still containing water.... .

Left - Notwithstanding the attention it will likely receive the coal is perhaps not of the best quality on No 49; notice too the spill on the cab roof and the electrification warning flashers on the tender - well this was the era of the Glasgow 'Blue Trains'.

This page - Scotland is not the only place to be associated with traditional 'winter weather', the Settle & Carlisle line similarly affected. Here Paul has braved the elements at Dent station to record an almost complete 'white-out' for the engine and crew.

Still keeping with the theme of weather I doubt there were many who considered this type of photograph, and yet it works well. The rain soaked platform (from unrelated records the location is thought to be on the Somerset & Dorset line (which we know Paul visited at some point), the interior carriage blinds and we even surmise Paul was alone in the compartment; else there may well have been much condensation on the inside.

Back to winter and conveniently with the station name visible. Braughing was on the Buntingford branch line in Hertfordshire and saw its first trains in 1863. A century and a year later in 1964 it witnessed its last. Serving the village of the same name it is possibly best remembered when it featured in the 1954 film 'Happy Ever After' with the station renamed Raithbarney for the picture.

We visited the Forest of Dean briefly in the first book and it is with pleasure that some more from that area and time can now be included. Previously the engine was No 3759, this time identification is obscured; perhaps the smokebox number plate has been removed as a souvenir. On the left the two views show a Pannier tank working relatively hard but below it is most definitely flat out and probably with the sanding gear operating to maintain grip. Paul's images are indeed wonderful but what a pity sound was not captured as well.

Two types of Southern 0-6-0, that on the left originating on the South Eastern & Chatham Railway and to the right a Drummond '700' class engine from the London & South Western.

These engines were some of the workhorses of the fleet, spending much of their time on freight but sometimes seen on light branch passenger duties.

On the left C class No 31280 has steam to spare and is engaged in shunting at an unknown yard; notice on the left the dilapidated grounded wagon body and the three-way point in the background.

Conversely No 30315 is at Eastleigh, both it and the 2-6-2T behind on the servicing road and with both also displaying the route code discs for 'light engines from stations west of Basingstoke to Eastleigh'.

Sign of the times as well with the motorcycles and similar combination on the right; an improvement on the pedal cycle although widespread car ownership was still a little way off.

Neither engine would survive into the future both ending their days in 1962.

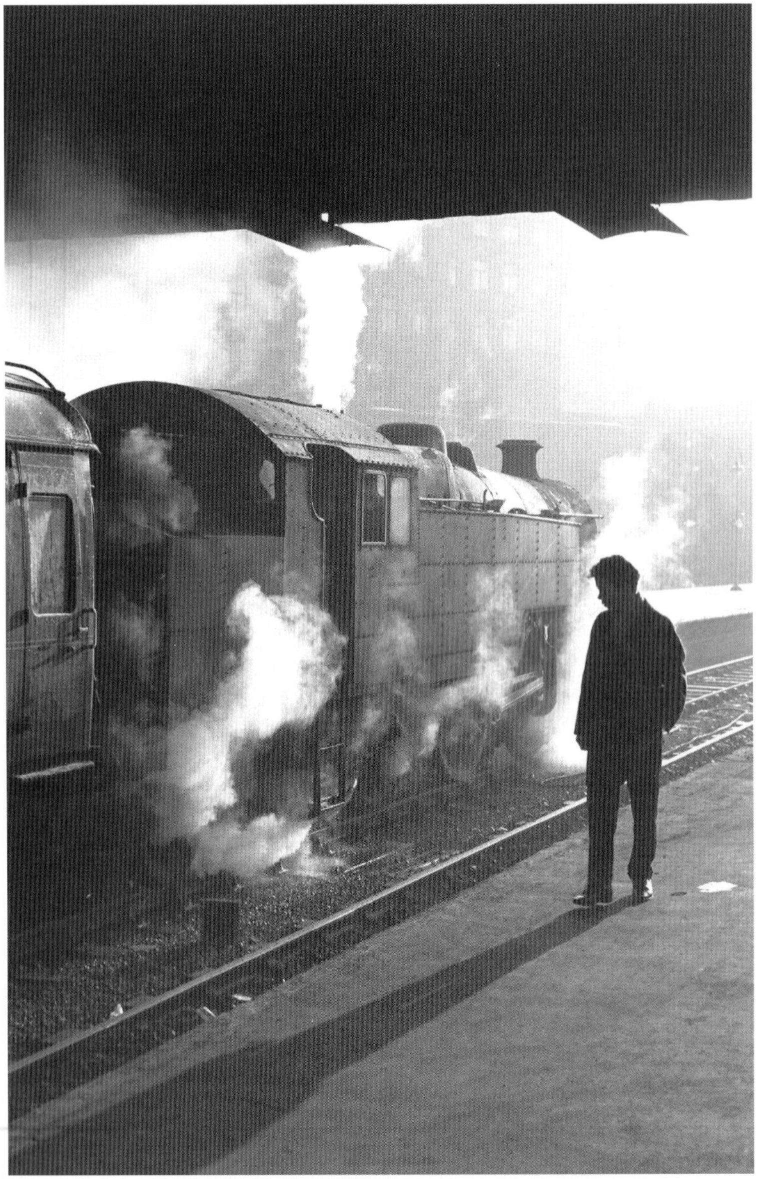

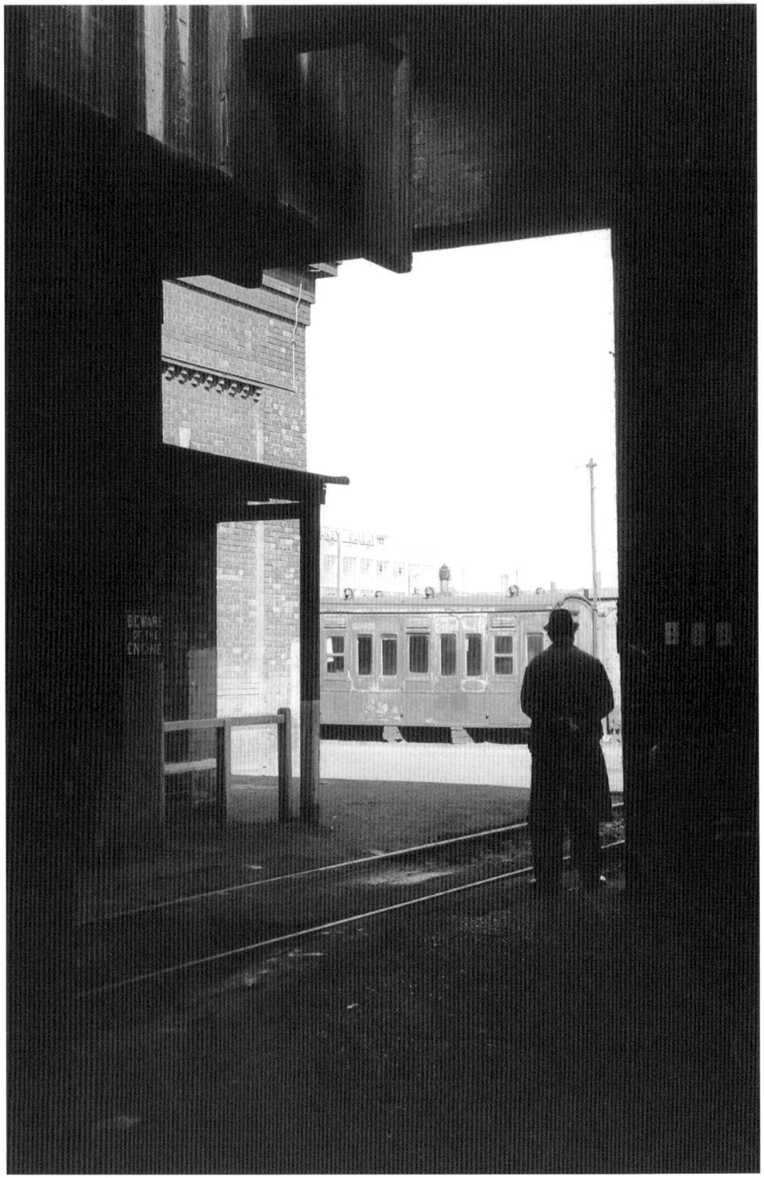

Observation times! All may well be railwayman and all seemingly engaged in deep thought. Locations on the left are not reported, above is Havant, No 32678 replenishing its tanks ready for a Hayling Island train.

Holiday time on the Isle of Wight, again a location worth a second visit. We are at Ryde St Johns on the left - the locomotive works are in the background - O2 No 18 *Ningwood* in the down platform probably on a shunt move within station limits.

Opposite the same engine is seen but this time at Ryde Pierhead taking water before heading south.

The bark from one these little engines working hard was not to be forgotten; your writer recalls a childhood visit to Cowes where the family were intent on viewing the yachts yet one small boy was captivated by the sound of engines hidden from view by the frontage of Cowes station. Grown - ups won (of course) and he never did get to see the trains at that place; but he did walk through the by then trackless tunnel at nearby Mill Hill several times a few years later... .

A spread of opposites.

Left we have two ladies in summer frocks although perhaps more of interest from the fashions of the day and the weather is No 20 *Shanklin* in the background. The location is Ryde Pierhead, the foot passengers seeming to prefer to walk to the ferry rather than use the train or the pier tramway.

On this side a day when summer clothes would definitely not be welcome. This is the water tower at what had been the locomotive shed at Newport and which continued to be available to engines even though the shed itself had closed in the late 1950s with all maintenance now concentrated at Ryde.

Extreme cold such as this is unusual on the Island which often serves as a micro-climate compared with the mainland.

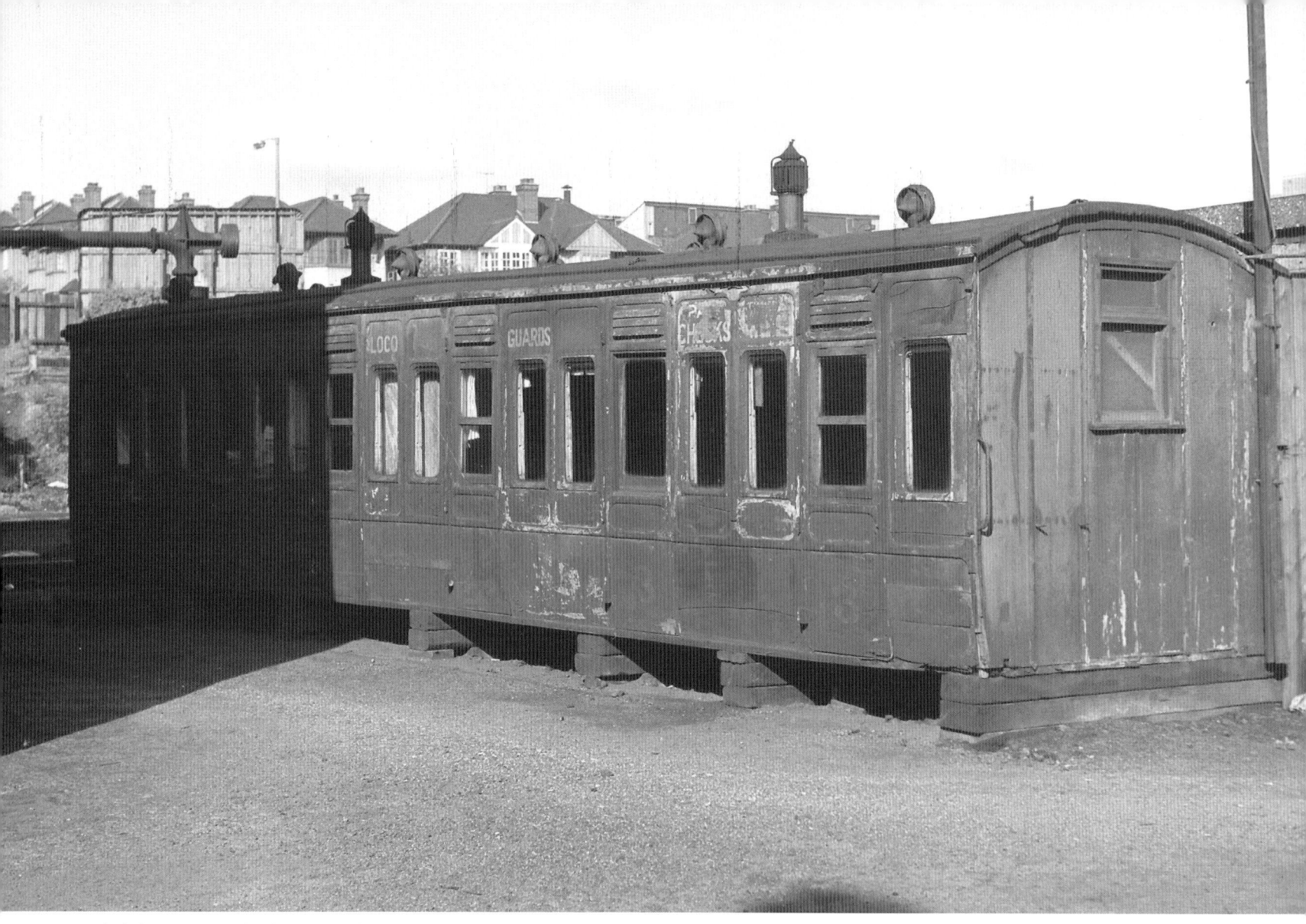

A slightly tongue in cheek pairing; the grounded coach body in use as office / messing accommodation and the station notice from yesteryear referring to a horse drawn carriage - and possible later the 'horseless carriage' (aka the motor car).

CARRIAGE ENTRANCE ON UP SIDE

Another pairing although this time of opposites. A brace of A4's, 60011 *Empire of India* and No 60020 *Guillemot* seemingly out of service and stored.

A4 replacement, Deltic No D9007 *Pinza* in the attractive original two-tone green livery with white window surrounds which so suited the class - complemented by silver-painted buffers. Disliked by the enthusiast fraternity at first - which one of us really does embrace change (?) - they would later become firm favourites and yet were destined to have a shorter life on front line duties than their steam predecessors.

30

Steam may have been displaced from top link passenger duties but for freight it remained dominant. In consequence servicing facilities remained, some becoming ever more decrepit, what money there was devoted to new depots for the new traction. Maintenance in the open in harsh and dirty conditions was unlikely to attract new staff whilst a further consequence of the change over were lines of stored engines identified as per the sacking over the chimney. Few would ever steam again.

With the prestige services turned over to diesel - or electric - the former top-link steam engines were often the first to go; intended for express passenger work they did not sit comfortably with freight or lesser turns. In consequence the year 1962 witnessed the end of the King class on the Western Region, No 6011 *King James I* forlorn at Swindon and already separated from its tender.

Also in store at Swindon were spare boilers, that on the left identified as a 'S' type as fitted to the 48xx (14xx) and 58xx 0-4-2T classes. A 4-wheel bogie from a 4-6-0 of some sort also sits in the foreground.

With Western diesels under construction in the background, some steam overhauls continued, No 7920 *Coney Hall* seemingly in the last stages of repair at Swindon. Undoubtedly this would be its final visit to 'the factory'.

The exception to the rule relative to the principal steam classes was on the Southern Region where formal approval for an extension of the electrified network was not given until 1964 and steam would remain supreme on the lines out of Waterloo until July 1967. This is No 35017 *Belgian Marine* beautifully turned out from Eastleigh Works after overhaul together with a Class 4 standard tank, No 80150.

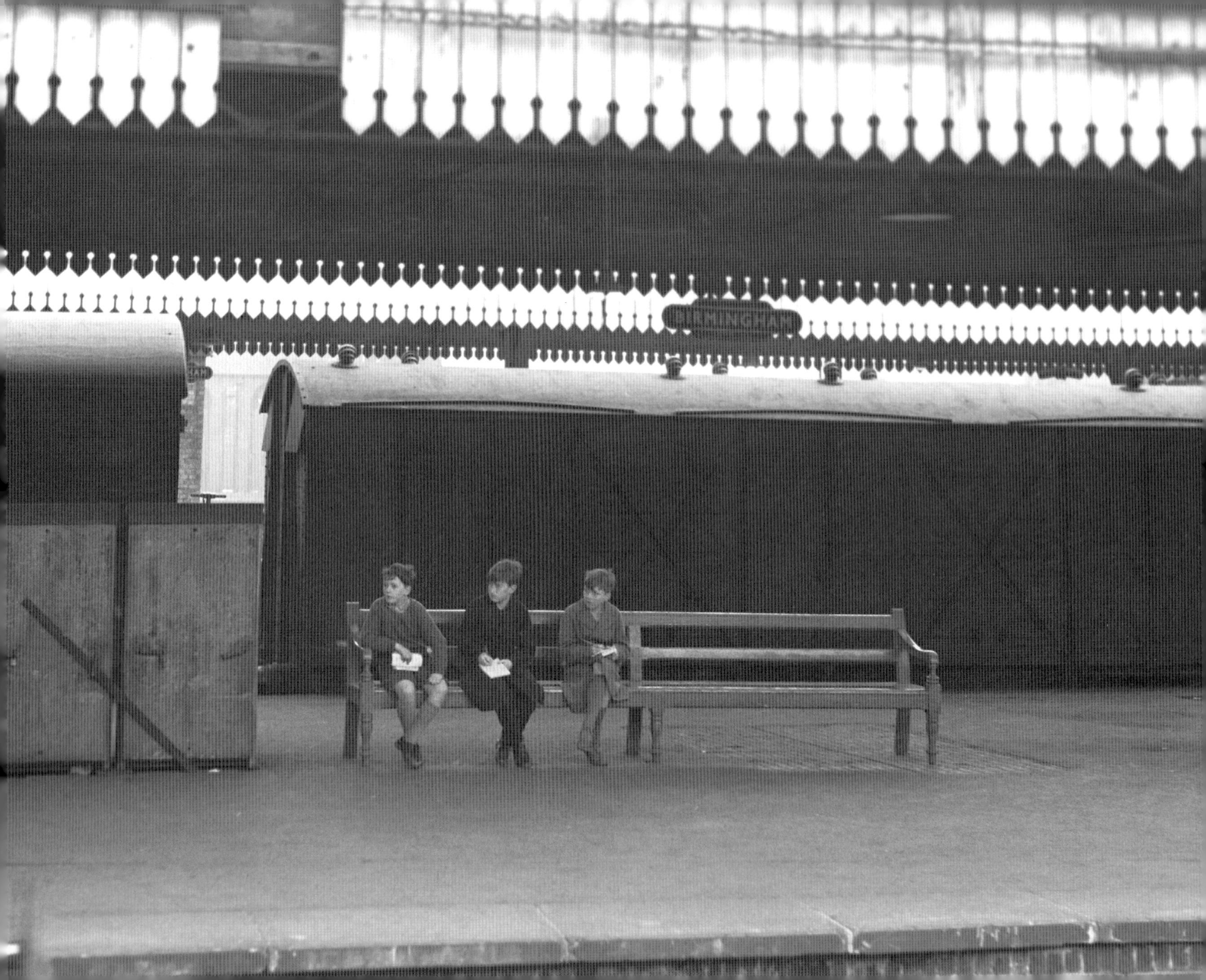

Left - A return visit to Snow Hill where two have also become three. So what is it that might be capturing their attention? With just a hint of a buffer coming into view it is an answer we may never really know although perhaps the best guess is a DMU.
Above - Two four-cylinder products from Swindon with both carrying the '81A' shedplate for Old Oak Common. On the left is an unidentified King; the squat boiler making the class readily identifiable from most angles. It carries the 'Inter City' headboard, a named service which operated between Paddington and Wolverhampton Low Level until 1965. Alongside is Castle No 5056 *Earl of Powis* on another Class A, but unidentified working.

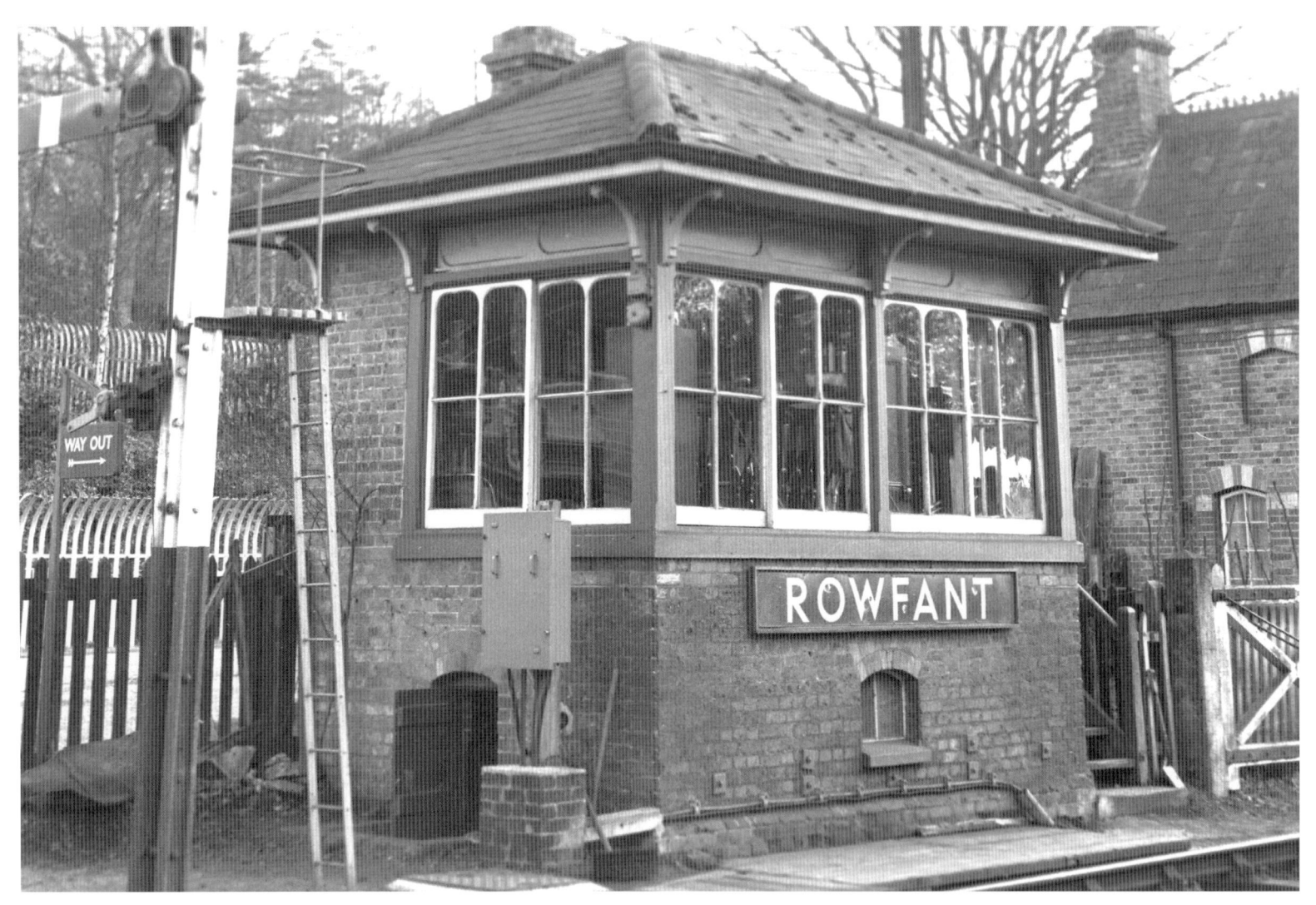

There were once in the order of 20,000 signal boxes controlling trains in their immediate area and between neighbouring boxes; today that number is probably little more than three figures and small wayside structures such as here in Sussex are no more. Rowfant had been a crossing place on the line from Three Bridges to East Grinstead, its resident signalman probably duplicating his role with that of issuing tickets and general portering duties. The brick signal box looks sturdy and has stood the test of time well, not so perhaps the signal post leaning at a jaunty angle.

Most signalmen would display a distinct pride in their work place; identified by a gleaming appearance with lever tops burnished, wooden cases polished and distinct shine to the floor - there was an all pervading smell of polish, Brasso and paraffin - once experienced never forgotten. Sunday was the usual day for 'housework', fewer trains and so time to clean and tidy ready for the week ahead. If a box had several men sharing the shifts then it was the responsibility of whoever was on duty for that particular turn. Hated were the relief men, who might arrive in dirty boots leaving their trade mark footprints within and who would then depart without making amends. Where a box looked untidy inside this was invariably because there were no regular staff and hence no one to take pride in the job.

Of course it could also happen that one of the regular men was not prepared to undertake his share in which case a tide mark would result; the regular man cleaning only half of the box and leaving the remainder untouched.

Paul Hocquard's view shows paraffin handlamps inside of a signal box we think somewhere in East Anglia. Lamps were sometimes cleaned and refilled in a separate hut outside but here that task is undertaken within the box. At night at least one oil lamp was always lit with the red glass turned so that if it were needed in emergency - and this would invariably be to show a red light - then all was prepared.

Signal box interior, possibly Princes Risborough or Brill & Ludgershall; we know Paul visited both. Continuing on from the description on the previous page, here we see the shiny floor, smart signalman, and lever cloth. When levers were pulled the cloth prevented grease from the hands tarnishing the lever tops. Although a solitary role (excepting in the largest and busiest boxes) there was a distinct pride in the job, a comforting environment in which to work and also one where it was on occasions necessary to exercise the mind in the regulation of trains so as to minimise delays.

'Out on the line' we have examples of GWR and BR(W) signals from different eras. Left is the earlier wooden post and bracket; the main arm for the connection from the loop on to the main line - a passenger line as well as there is the cover of a facing point lock present in the 'four-foot', the other shorter arm signal on the bracket is for what may well be a goods loop. To the right is a later tubular metal post for the main line. We may conclude also the controlling signal box is some way behind as there is a telephone box at the base of the main line signal.

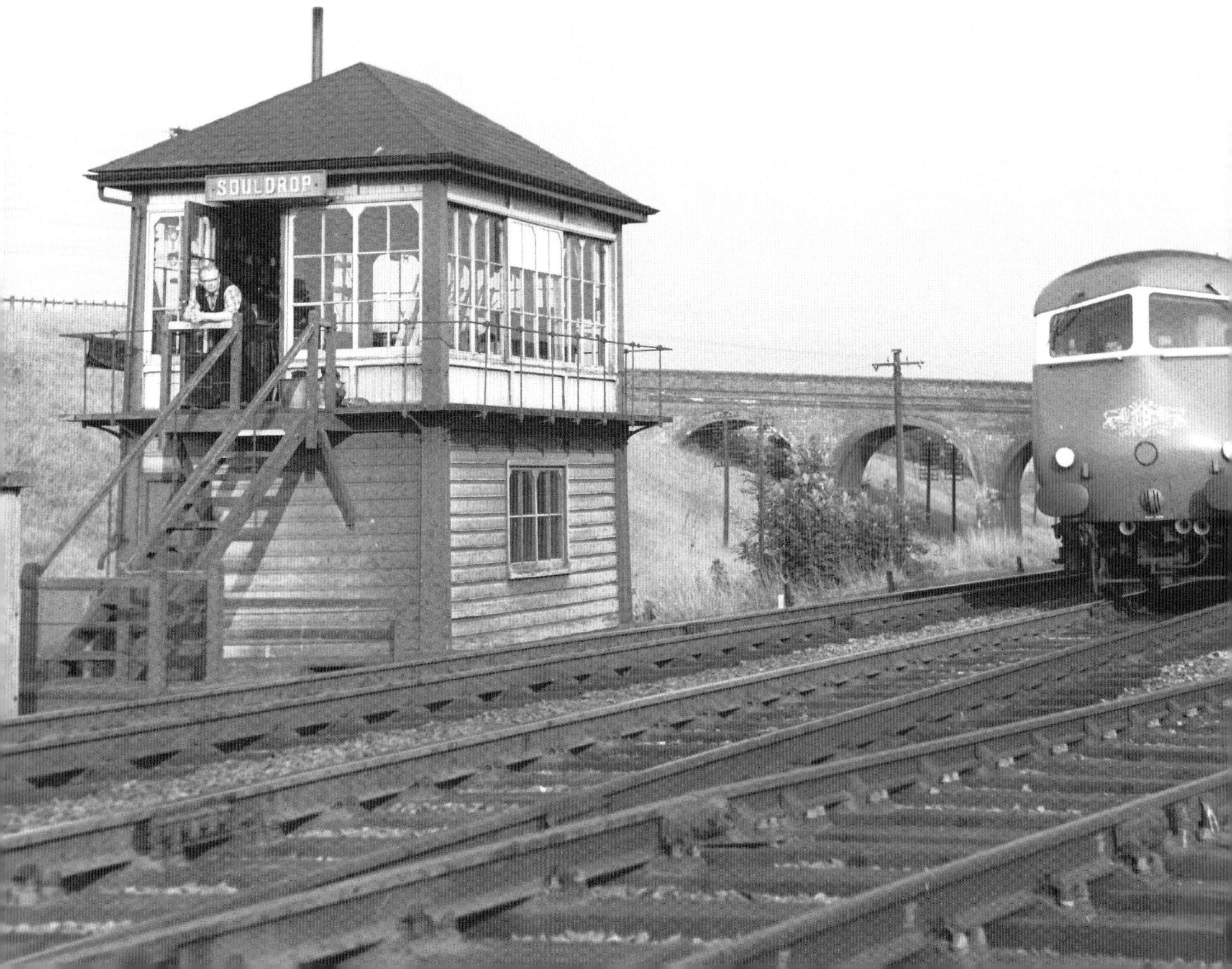
SOULDROP

Left - The prestige 'Midland Pullman' train passing Souldrop signal box, Bedfordshire. This train, introduced in 1960, operated daily between London St Pancras and Manchester Central for just six years until 1966 providing first-class accommodation only aimed at the business traveller. Similar services operated on the Western Region but this time catering for both first and second class - although still Pullman. Considered by some to be the forerunner of the later HST sets, the connection between the two was more accurately the fixed formation service rather than the type of accommodation on offer.

Above - Where signals and signal boxes go to die. A commonplace and melancholy sight up and down the land in the 1960s. Redundant mechanical signalling either cast aside when lines and stations closed, or in consequence of a modern 'multiple aspect' (colour-light) scheme. Signal boxes too were no longer required, most simply demolished as no longer serving any useful purpose although a few were retained and converted either into residential accommodation or continuing in their original role on the network of heritage lines up and down the country.

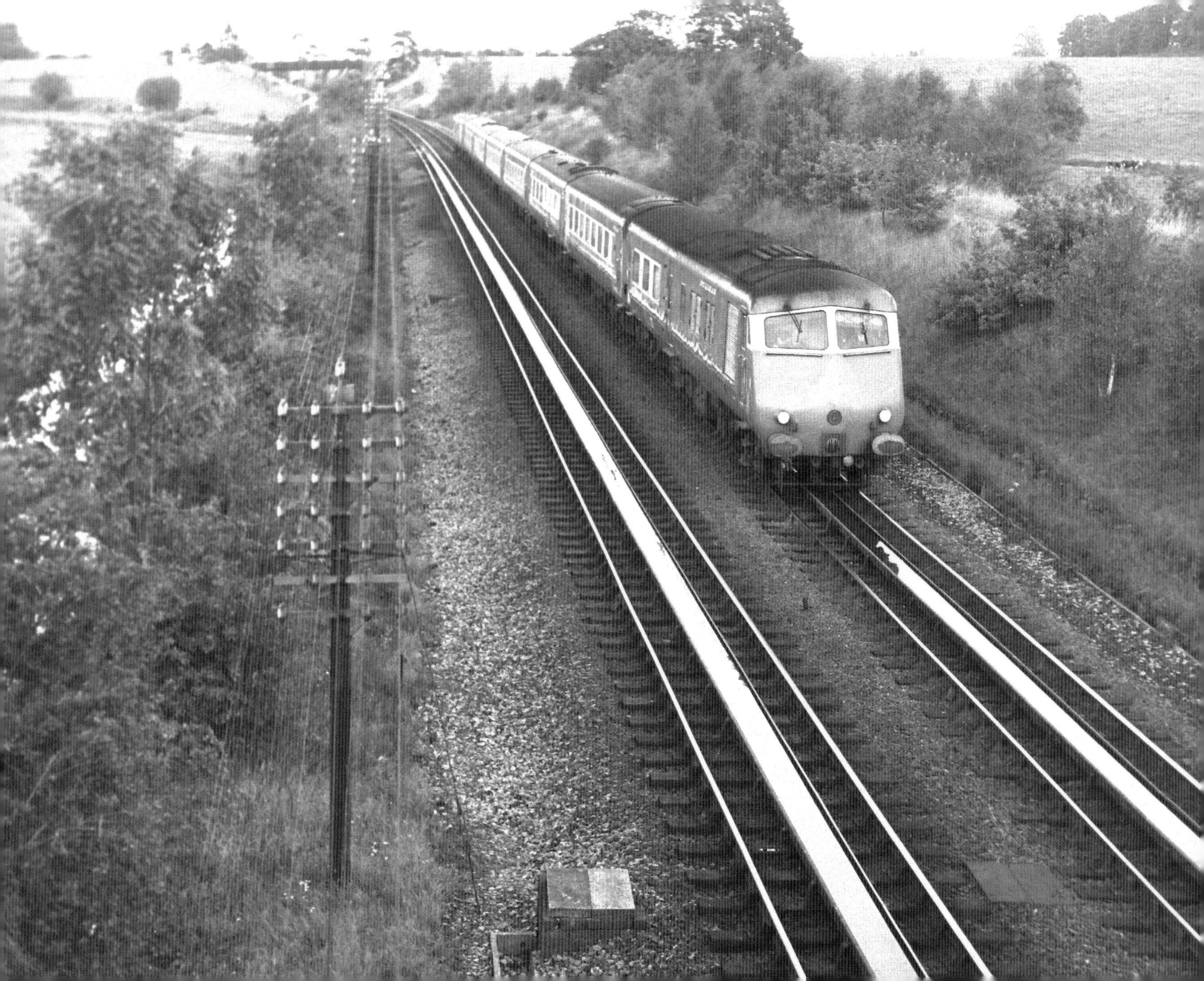

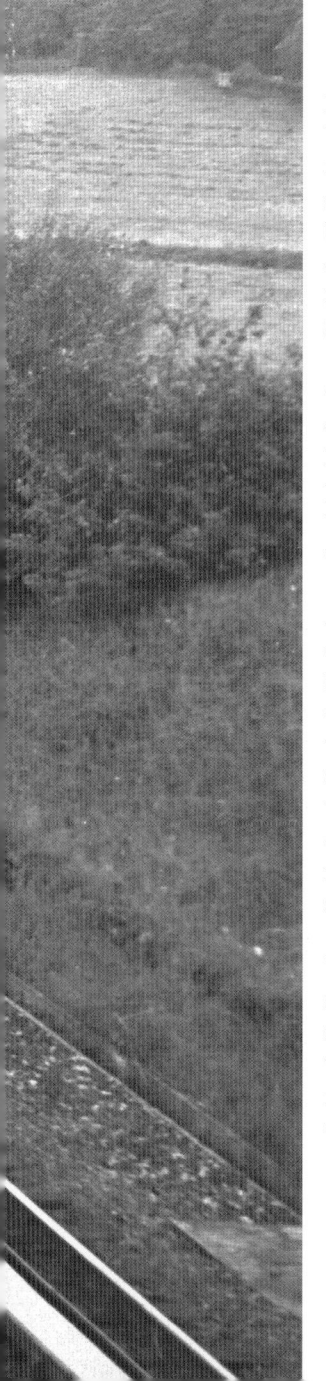

More on the Pullman services, the Western Region's train running as 8-coach sets and initially affording a daily service in each direction between Paddington and Bristol and Paddington and Birmingham Snow Hill. (Later a Paddington - Swansea service was added.)

Operating on diesel power with a motor coach at either end, 2,000hp was available for traction purposes, two auxiliary engine units providing power for air conditioning throughout.

Although popular so far as the on-board service was concerned - a meal was available to be served at every seat (at extra cost of course) without the need to move to a separate catering / restaurant car, the ride was not good excepting on the very best track; in which respect the Western Region fared slightly better than the Midland.

Unfortunately with all three WR sets in regular weekday use, when a failure did occur or a works' visit was scheduled, recourse had to be made to a locomotive hauled set of old fashioned vehicles.

The modern(ising) railway. Sulzer Type 2 No D5036 in charge of a parcels working, the two headcode discs identical to the steam era and meaning the train was 'fully fitted' (having the continuous brake). The engine appears to be possibly freshly into service in smart green livery with a yellow warning panel on the front end. A corridor connection is also available at the front should two locomotives be coupled in multiple and thus allowed access between the two for the purpose of attending to faults etc. In practice these corridors were rarely used and were later sealed due to the ingress of draughts.

Emerging from a tunnel (thought to be White House Farm near High Wycombe), a three coach set has the duty of a semi-fast service, Again green livery with what were referred to as 'speed whiskers' at the front; to act as a warning to anyone working on or in the vicinity of the track.

A pair of Cravens built DMU sets at Kentish Town. Compared with steam, passengers welcomed the open design of most of the DMUs, forward vision also possible often for the first time as well - although some drivers did not particularly like being observed and would pull down the night blind behind them.

The former Great Eastern lines in East Anglia were ripe for modernisation and a number of the early 'First Generation' as they are now referred to, DMU sets were sent to the area. Although they succeeded in pausing the decline in passenger numbers routes in the area were experiencing, it was not sufficient to increase revenue and Long Melford was destined to be one of many to lose its rail facilities completely. (To the right the signal for the line to Lavenham and Bury St Edmunds has already been removed and the track lifted.)

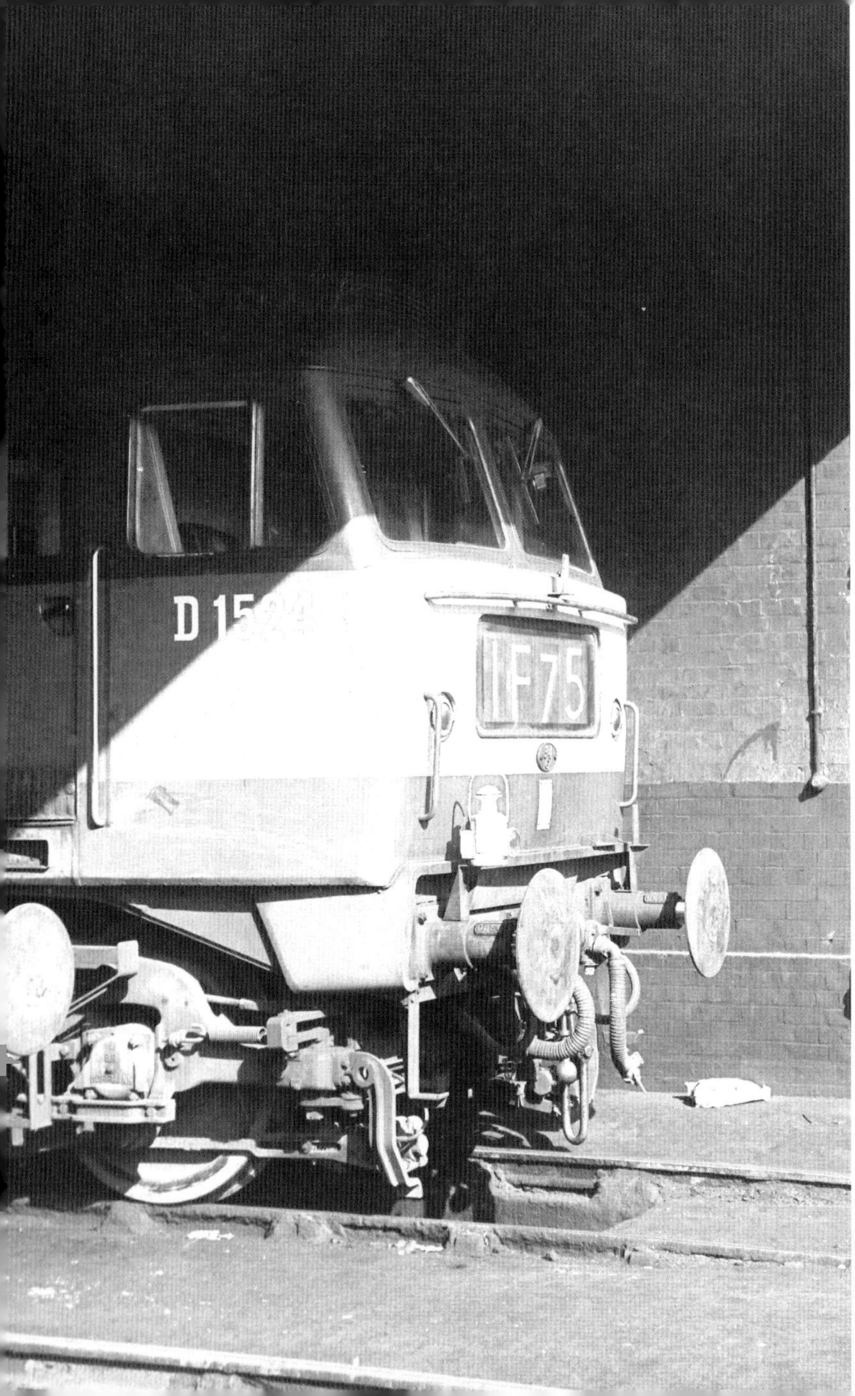
D 15
1F75

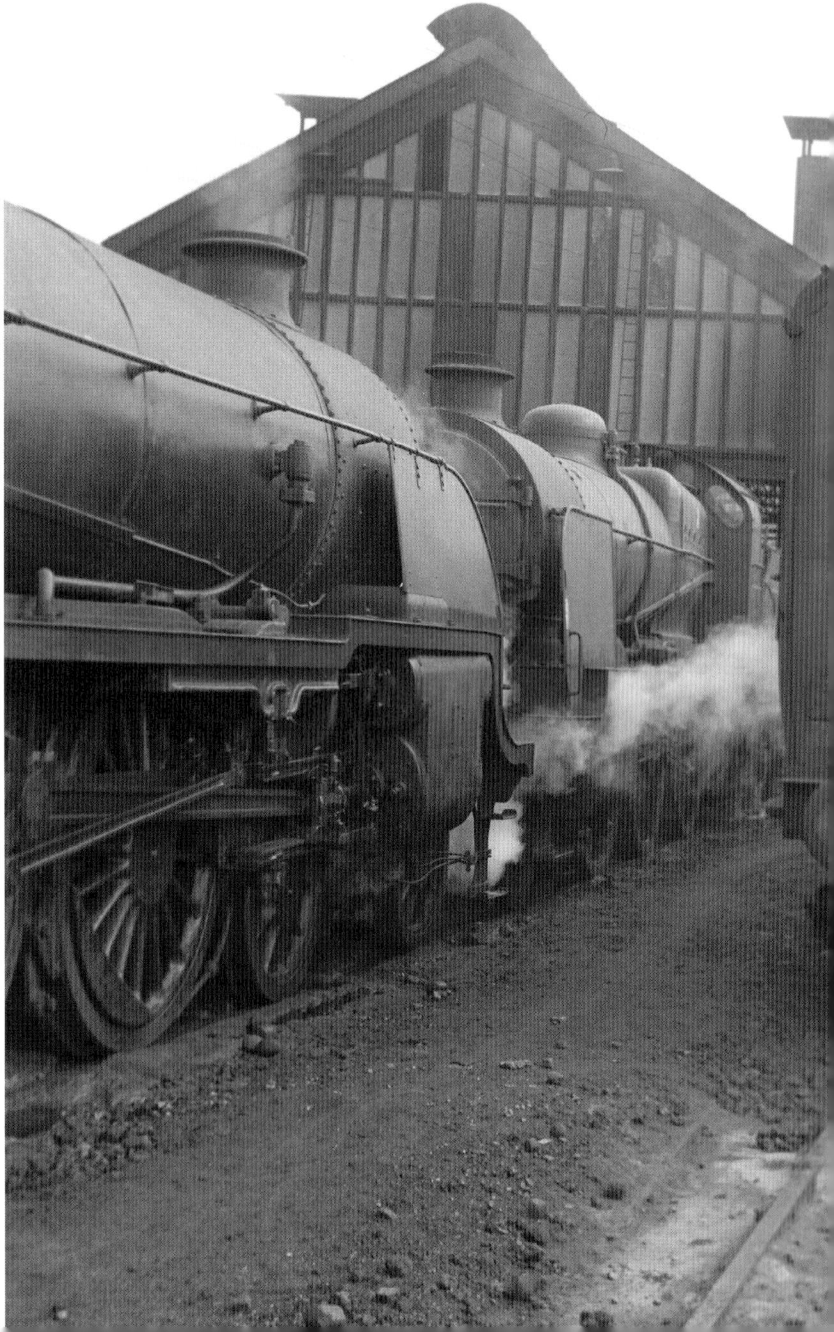

In some locations, new diesels might be seen sharing their stabling points with older steam engines - why the words 'stabling point?' - well simply explained as dating back to the earliest days of railways when the steam engine was referred to as the 'iron horse' and of course horses were kept in stables. Consequently we see a new Brush Type 4 within a former steam shed whilst elsewhere a pair of Southern engines are outside a smaller depot and an A3 and others completing the picture at a larger depot.

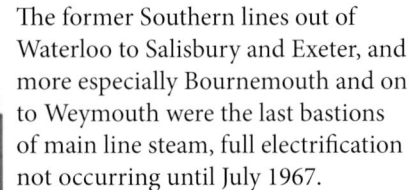

The former Southern lines out of Waterloo to Salisbury and Exeter, and more especially Bournemouth and on to Weymouth were the last bastions of main line steam, full electrification not occurring until July 1967.

Before that time there had been a steam depot at Basingstoke where S15 4-6-0 No 30498 was recorded seemingly waiting for the ground disc signal to clear after coming off-shed. On the right is the somewhat primitive coaling facility.

A few miles south No 35008 *Orient Line* in appalling external condition is at Micheldever with a down Bournemouth train. We were at Micheldever at the start of this book, the signal box on the left also just visible in the background to the previous station views.

Finally we have No 34066 *Spitfire* once more in typical late Southern Region dirty condition on westbound milk empties at Battledown, taking the Salisbury / Exeter line. Milk from stations served by the Southern lines was a regular daily traffic and of course 'what went up (to London) must also come down'.

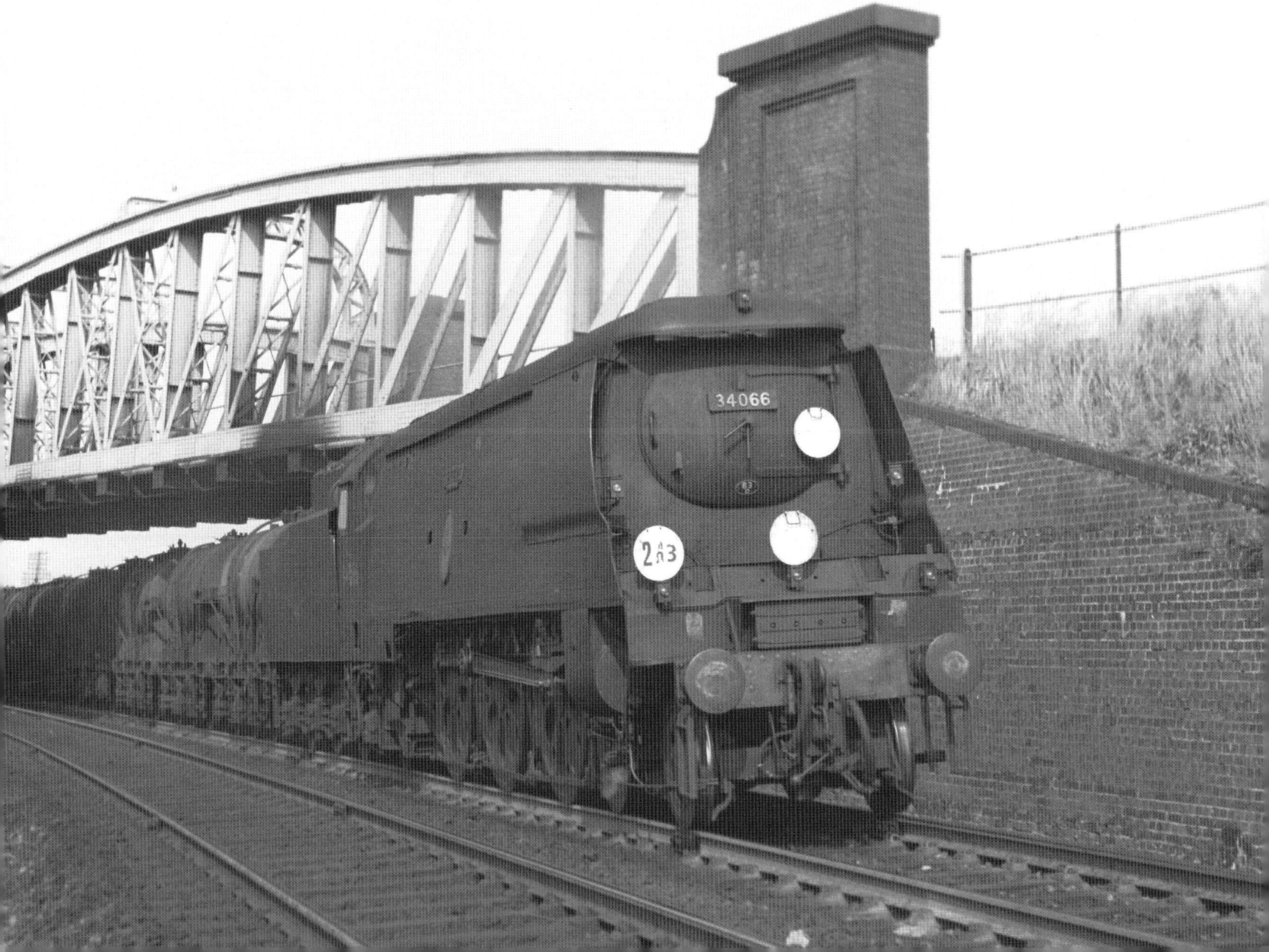
34066

No 35008 again, possibly on the embankments east of Basingstoke. Rebuilt from their original 'air-smoothed' condition, the revised design was intended to retain the best of the old but without the erratic behaviour. It must be said the redesign, attributable to R G Jarvis, was a success, externally the engines now bearing a distinct family likeness to the BR Standard pacifics whilst still retaining an amount of Bulleid 'charisma'.

The short branch line from Brockenhurst to Lymington survived the cuts of the 1960s possibly because of its connection to the Isle of Wight ferry serving the western side of the Island. It would also gain the distinction of being the last steam worked branch line anywhere in the country although in 1967 motive power was limited to BR Standard and Ivatt tank classes, the M7 design seen here crossing between the Town and Pier stations at Lymington having disappeared in 1964.

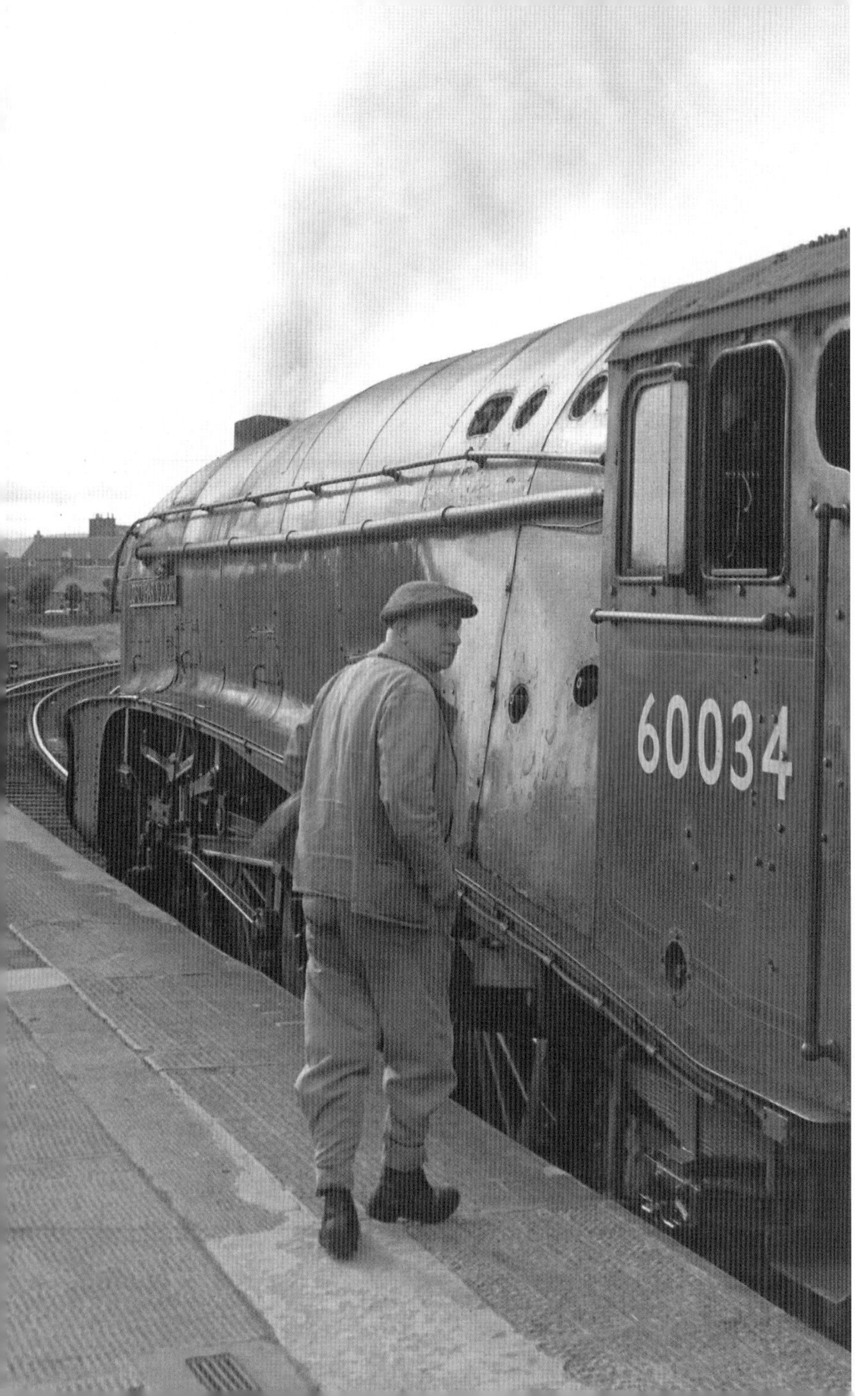
60034

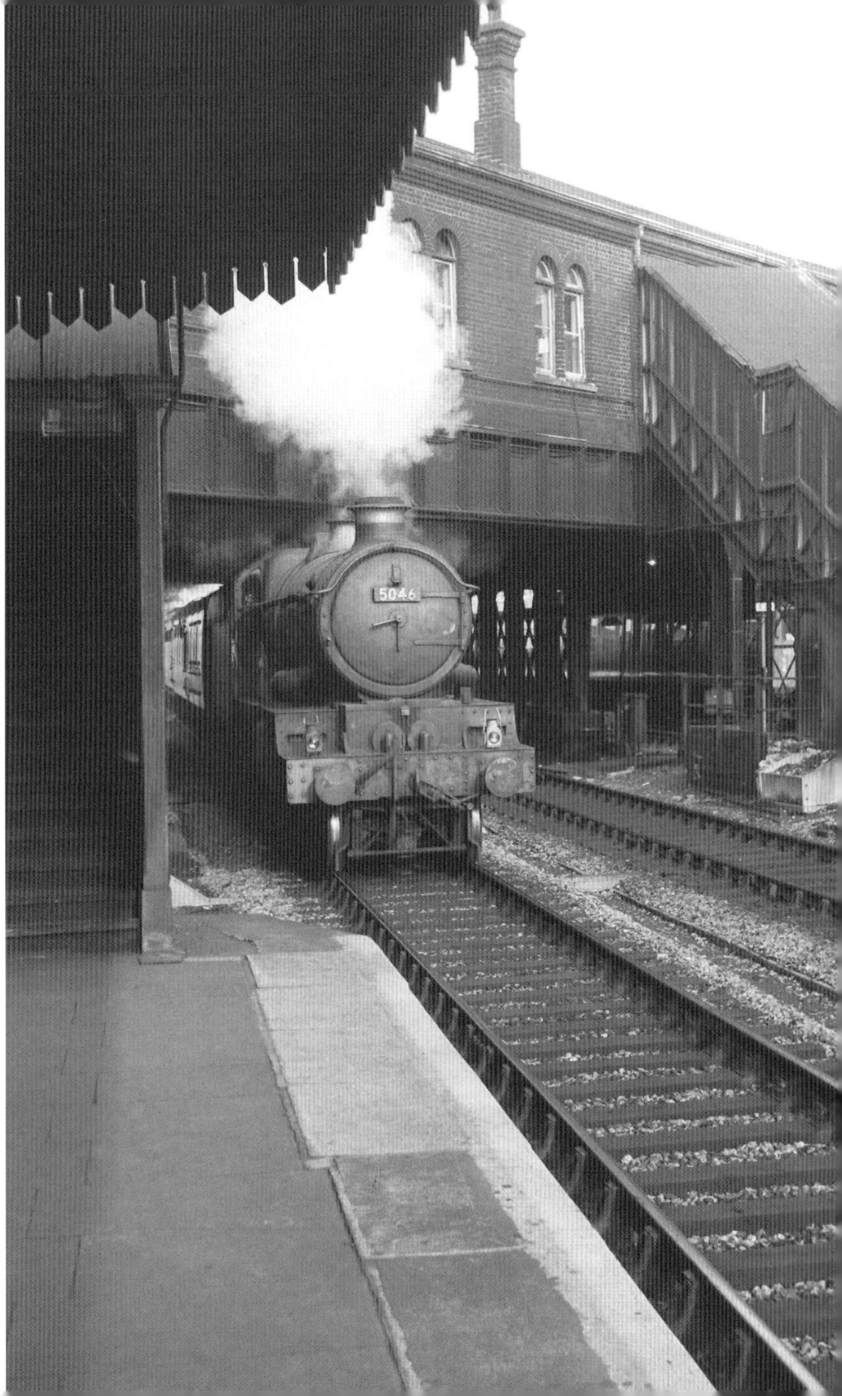
5046

Main line steam, Nos 60034 *Lord Faringdon* and 5046 *Earl Cawdor* on their respective top link duties. And yet 'they also serve those who stand and wait', Pannier No 9405 on pilot duty at Paddington

Changing times - changing needs.

Top left - Once upon a time there were hundreds if not thousands of railway horses, their role to shunt wagons and pull drays delivering and collecting goods. On the road they were replaced by the 'mechanical horse', for shunting, it was either by an engine or simply because the need no longer existed. This was 'Charlie', BR's last shunting horse at Newmarket.

Bottom left - On the ground a pair of wagon turntables. Installed in goods yards at a time when a wagon might be turned manually - or with the aid of a horse. Dating back to the 19th century they would somehow not fit in with the modern privatised railway.

Right - Inside the cathedral like expense of the station, a single wagon - 'tail traffic' - as it might be called; ready to be added to the tail end (the last vehicle) of a train - and nothing to do with horse boxes although a horse box could indeed still be tail traffic. Special instructions were issued by the railway when trains like the 'Midland Pullman' came upon the scene, this was that 'tail traffic' was not to be added to the service.

Far right - A signal box closed, the signalman likely made redundant and in its place a simple open air ground frame to control the remaining points and sidings. No more warm stove, no armchair, and no polished lever tops.

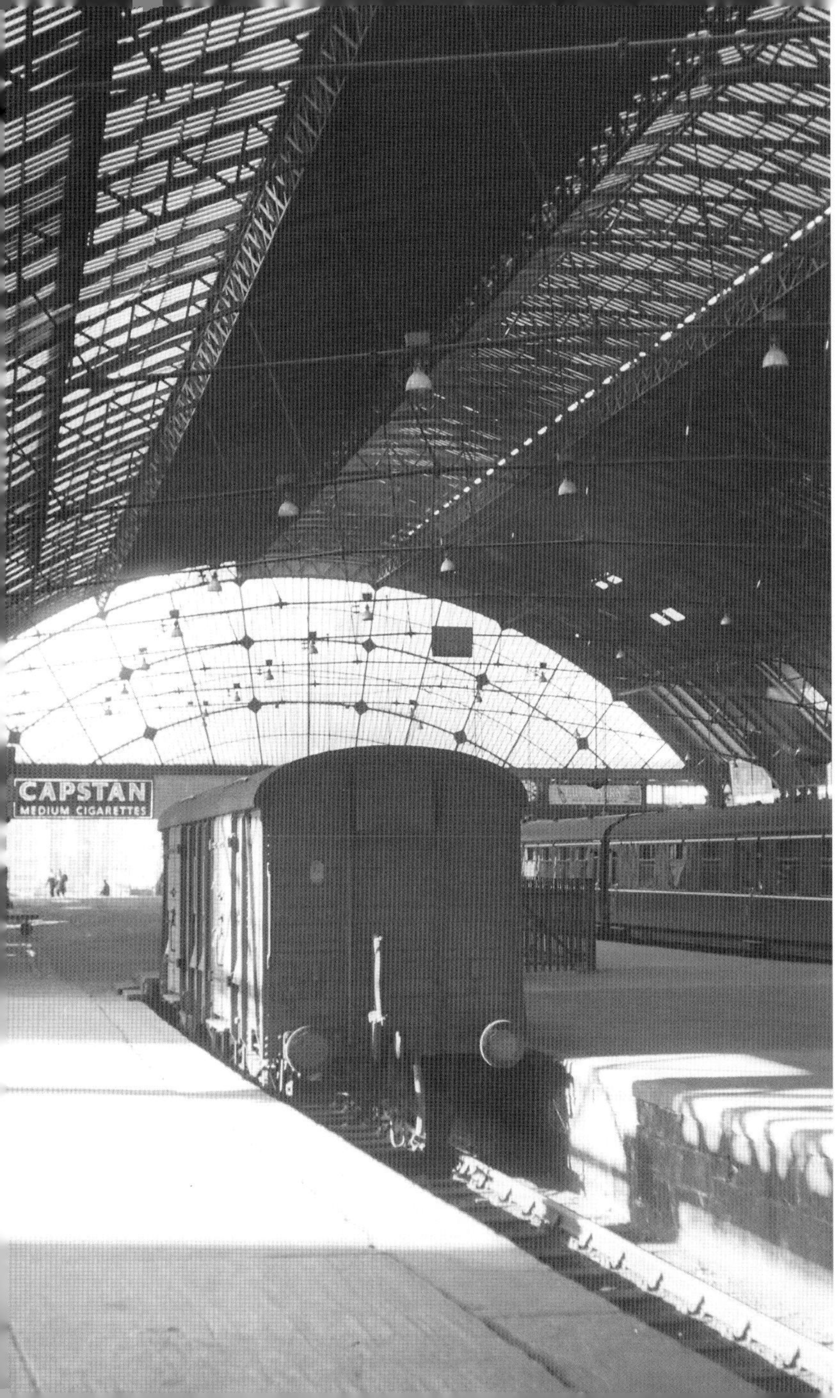
CAPSTAN
MEDIUM CIGARETTES

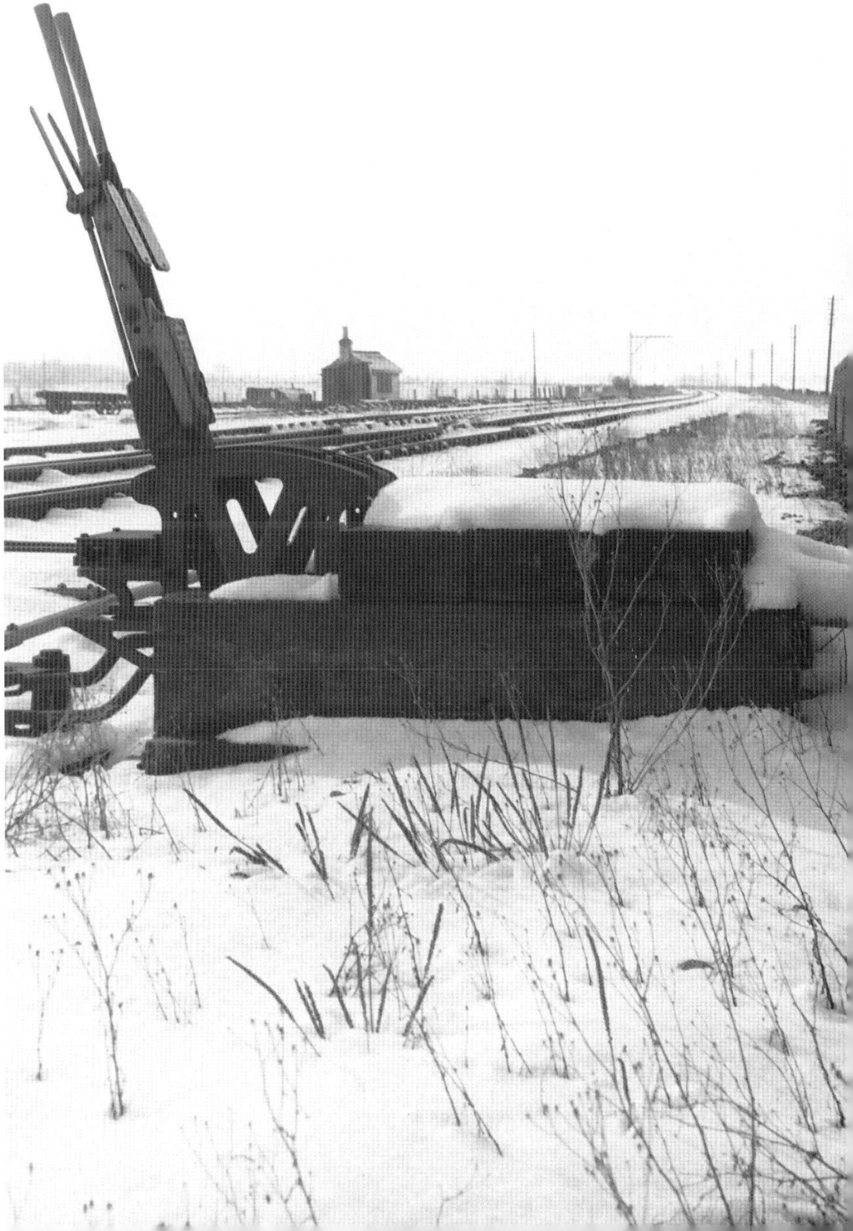

Redundant, or soon to be redundant assets.

Left - A ladder crossing; so called as it crossed several lines similar to the rungs of a ladder. Maintenance for such a crossing was considered high and in consequence few now survive their place taken by a series of simple crossovers; more space but less cost.

Above - Perhaps an unintended closure. The Severn Bridge railway closed as a result of a barge colliding in fog with one of the bridge piers so bringing down a girder. Had this unintended accident not occurred, who knows, might it have been a survivor?

Right - Mothballed pending either the demolition contractor or a revival? In reality few lines that were closed were restored - certainly not under public ownership. This example is on the Hawkhurst branch in East Sussex. Demolition came soon after.

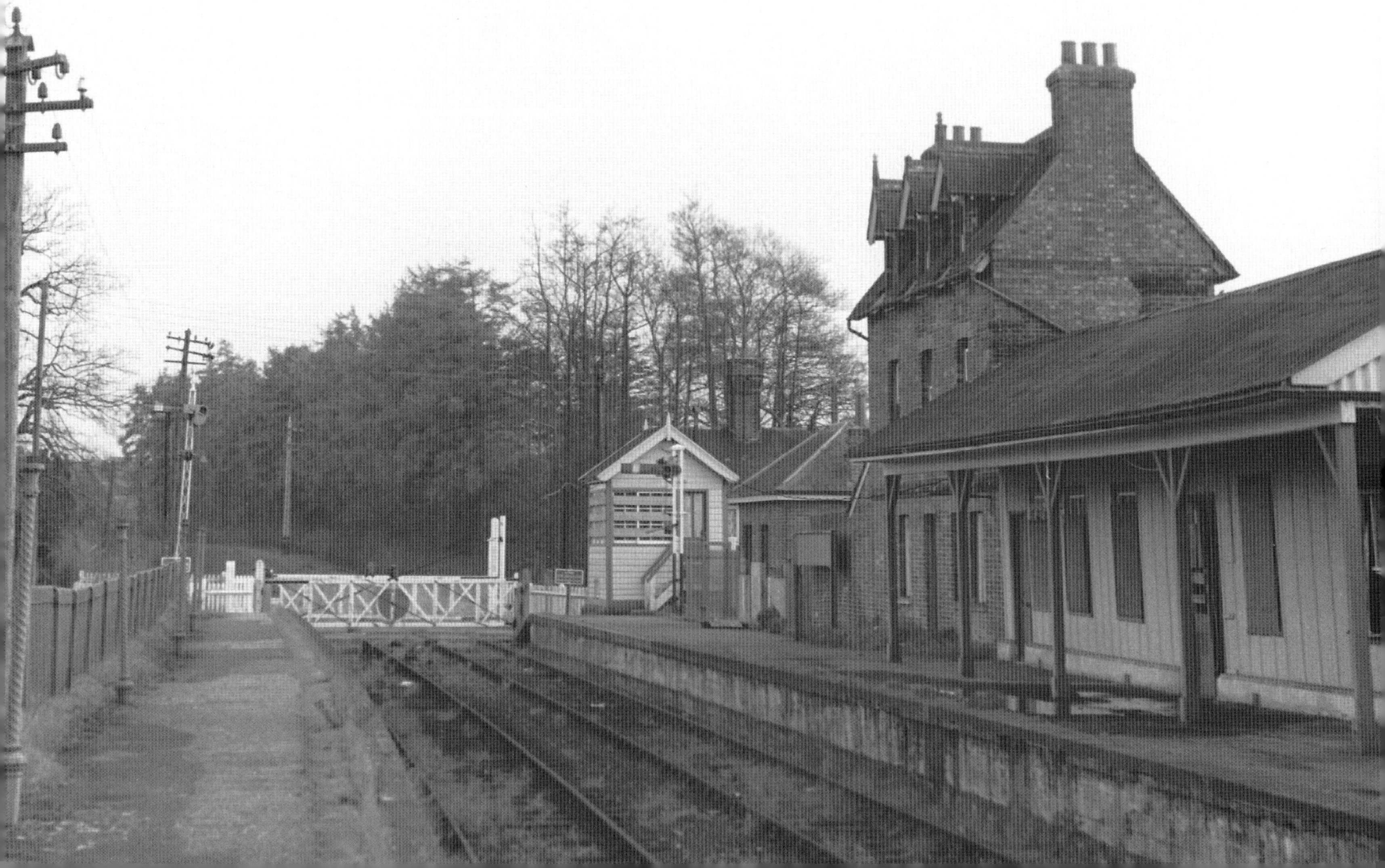

Sometimes a location could remain almost in a state of limbo for decades. Such was at Bishops Waltham in rural Hampshire where the little terminus had lost its passenger service back in the 1930s and yet the line remained for goods, much of it household coal, until the mid 1960s. Soon after the photograph was taken all was swept away, the location is now a roundabout.

Another rural location, Welford Park in Berkshire, the mid-way point on the 12 mile branch between Newbury and Lambourn and the only stopping place on the branch with a passing loop. Closed to passengers in 1960, it somehow managed to survive until 1973 not least due to the presence of a private siding to a nearly military base.

The terminus at Lambourn with the railway now redundant - what price the signal box nameboard? Although closed prior to Beeching this was to be the fate of over seven thousand miles of railway in England, Scotland and Wales. Several of these closures now seen as a step too far but at the time considered necessary on the alter of economics and cost savings. To be fair Lambourn is not one of them.

Still at Lambourn, the goods shed dating back to when the line had been independent and possessed its own unique style of structures. Oh what stories this building could have told; initiation ceremonies for new staff (posted all over with luggage labels adhered by an evil smelling fish glue) and when used by the ladies of the area and a railway guard of particular repute… . Housing now occupies the site.

Inside the workshop. This is Swindon where a 'Teddy Bear' is under construction - identified as a D95xx shunting / trip locomotive. Why 'Teddy Bear'; folklore has it the works manager was walking through the works bemoaning the lack of orders and that this once proud establishment was reduced to the construction of shunting engines. He commented, 'We built the Great Bear (a reference to the steam 'Pacific' loco) and now we build 'Teddy Bears'.

Pride in the job. On the Guildford to Redhill line; carefully sculptured by a ganger of long past and still tenderly pruned by a later generation.

Amongst the major casualties of the railway closure programme of the 1960s was the erstwhile Somerset & Dorset line from Bournemouth to Bath. Never seemingly considered for more modern traction it remained a steam hauled railway until the end; an end that came in 1966 and locations such as here at Blandford Forum lost their railway lifeline for ever.

Paul was one of a select band who did not confine themselves to one location, seeking out the rural railway before it disappeared for ever. Here we see his work in capturing the rear of a branch freight, and at right recording a fellow photographer on their own outing - perhaps even the pair worked hand in hand. A notable feature of the country lanes of the period was the almost complete absence of traffic.

The railway in town and country, or might it also be described on both pages as 'ancient and modern'.

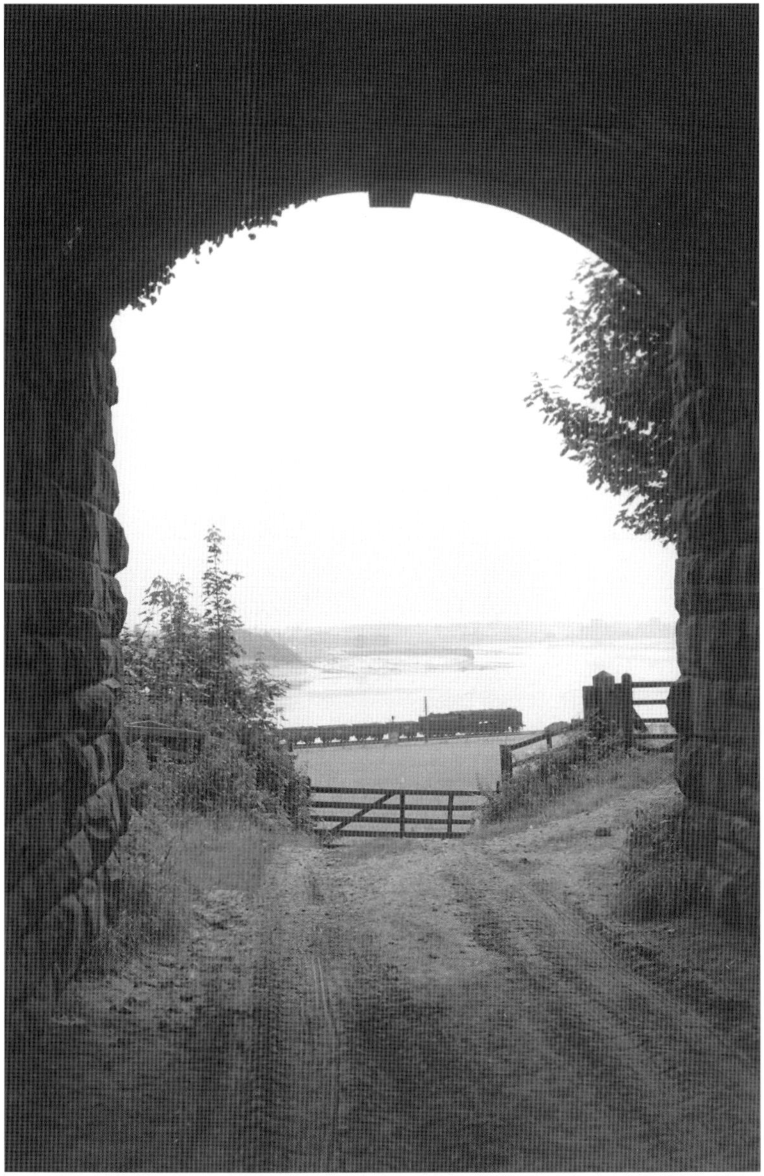

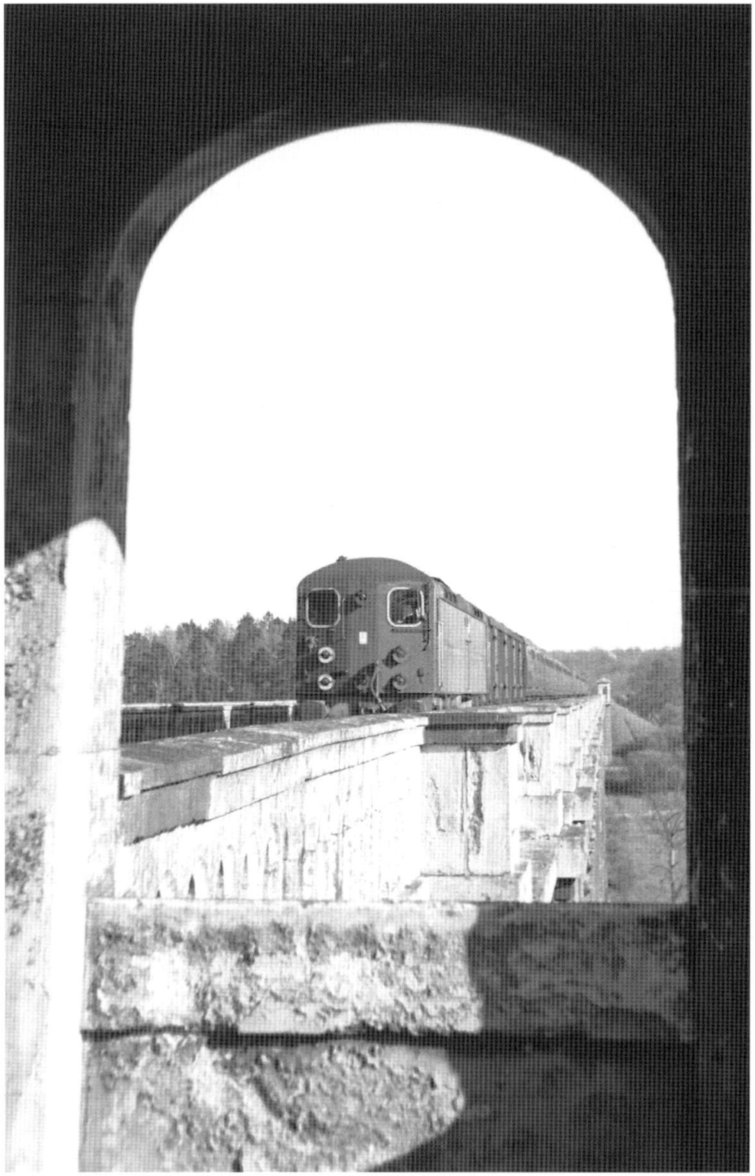

The BR Standard designs totally 999 new steam engines over 12 different classes. The Britannia type, represented here by No 70030 *William Wordsworth* first entered service in 1951 and at one time or other could be seen on all the regions of BR. The 9F type came three years later in 1954 and proved themselves to be arguably the best of all the Standard designs. They could though be hard on a fireman when worked to the limit.

71

Steam depot - 1 'weighbridge'. Here the weight of individual sets of wheels on axles might be measured and the springs adjusted as necessary

Steam depot - 2 'water'. Having been prepared for service the last jobs before leaving the depot are to trim the coal and take on water.

Steam depot - 3 'roundhouse'. Every engine accessible and having its own stall accessed via the central turntable.

Steam depot - 4 'ash pits'. A relatively clean steam depot, but in the background the new traction awaits.

76

Opposite - Conversation time, we know not what. Railwaymen's conversations were probably little different from those in other industries; work (complaining of course), wages (likewise), football (the referee got it wrong), and perhaps the plans for the weekend. Meanwhile those lamps are probably needed somewhere....

Opposite - reflections.

Top - Most steam depots also possessed a breakdown crane. Shown on the maker's plate was the notional capacity but this might often be exceeded in practice. This example was from Messrs Stothert and Pitt of Bath dating from 1909. Cranes were used not just for derailments but also to assist with civil engineering work. This particular example was photographed at Brighton.

Bottom - Former London & North Western Railway four wheeler tank of unknown purpose. The wooden brake blocks may be noted. Odd and old relics such as these might be seen hidden away at many depots and in yards.

77

Arguably the most important man at any shed was the one who yielded the broom, and carried a shovel in his wheelbarrow. His was the one who did his best to keep walkways clear and obstructions clear for the unwary. It was a thankless task especially at a big steam where there were comings and goings 24/7. How many too might be tempted to ignore the instruction 'Blowing down of boilers prohibited in this road' - perhaps it was outside the foreman's office!

Coal (and water). Coaling by hand at Hayling Island - a small engine only carried a small amount of coal and on a summer Saturday trains were full and a lot of coal was burned. For years such roles were simply seen as 'part of the job' but it was getting ever more difficult to recruit in the 1960s.

Meanwhile a diesel needs water too; for its train heat boiler (coaches were in the main steam heated well into the 1960s), and a steam era water column does the job perfectly well. The diesel is a modern 'Deltic'!

79

Departure time..... .